How Often Do You...

The Incomplete List of Thought Provoking Questions You Didn't Know You Needed

B. W. Daniels

Dedication

To everyone who has helped me along the way.

About the Author

B. W. Daniels has been writing since he was in elementary school. He graduated from Otterbein University with a degree in writing. This didn't suddenly turn him into an author, however. After years of only keeping work in notebooks he has decided to share some of his thoughts. He hopes you or someone you know will find the questions in the book helpful to your writing or just in finding out a little more about yourself.

think you can replace it?

listen to the same song?

wait until it's too late?

wonder about what should have been?

worry about what's coming up?

know it's all your fault?

wish it could be better?

wish it could go back to the way it was?

know you could do things differently, if you got another chance?

spend too much time worrying?

try to solve problems?

cause problems?

give up?

try again?

panic?

get up anyway?

say yes, when you want to say no?

want to try harder?

want to do better?

want to be better?

worry about everyone else first?

think about what you need, and not feel guilty?

take time…

    …for yourself?

    ...to read?

    ...to write?

    ...to build?

    ...to make?

    ...to play?

not tell someone what you need?

do something to make someone else happy?

not worry about the consequences?

ignore what you need?

tell someone what they mean to you?

tell someone how much you appreciate what they do?

let the moment pass you by?

not want to anymore?

say yes without thinking?

visit a city you've never been to?

think about walking away?

let yourself do nothing?

overreact?

under react?

not react?

think you should be doing something else?

let your mind wander?

let yourself wonder?

let someone in on the secret?

close the door?

let someone in?

write down what you are thinking?

hope someone will find what you wrote?

throw what you wrote away?

stare too long?

look in the wrong direction?

know you're right?

ignore what your body is telling you?

choose to set a good example?

clean out…

>…your car?

>...your desk?

>...your mind?

think about your biggest regret?

share?

give away your time?

remember the "good ol' days"?

realize you're in the middle of the "good ol' days"?

share what you wrote?

write in a specific way when you know someone will

read what you wrote?

act honestly?

know why you're here?

learn a new skill?

create something new?

work on something when you should be doing something else?

think something is more important than what you want?

think someone is more important than what you want?

want to stand up and scream?

feel like a fool?

want someone else to answer the call?

play games?

win?

lose?

believe the rules changed?

believe the rules only apply to you?

wonder how everyone else gets to do what they want?

know you're not who they think you are?

know absolutely nothing?

not know who you are?

not know where you are?

make stuff up, just to answer the question?

look at a blank page and have no idea where to start?

swear too much?

think about contacting someone from your past?

decide not to contact someone?

start writing before you realize who you're writing to?

want to answer that spam call?

wish you'd learned to play a musical instrument?

have to find something, for someone else, to do?

wish you could just stay in bed?

wish you were more brave?

wish the answer was yes?

want the day to be one more hour long?

want time to slow down?

want time to speed up?

want time to go backward?

ignore your work?

hear your song without choosing it from your playlist?

need your own soundtrack?

wish people would just get the point?

try something new?

return to what you know?

stand in the corner and hold your breath?

look at…

    …the floor?

    ...the ceiling?

    ...the stairs?

want there to be more?

want there to be less?

make time?

try to hide from the truth?

worry about repeating yourself?

worry about repeating yourself?

hit send before you finish your thought?

delete your thought before you send it?

wish you had something to write about?

start writing and you're not sure where to stop?

make it look like you're busy?

make plans and then find an excuse not to go?

scroll through social media?

buy something you don't really need?

re-invent yourself?

wish you knew all the answers?

wish you were allowed to do whatever you want?

think about not following the rules?

A safe space to write down some thoughts

think things would be better if you were Batman?

want to learn something new?

worry if you stop, you'll never start again?

think about the forces of nature?

realize you need re-assurances?

realize you haven't moved in hours?

put the pen down and look around?

put the phone down and look around?

put the phone down and talk to the person next to you?

know there's no one like the one person, your person?

admit good is good enough?

multi-task?

spend time stuck in the past?

write until the pen runs dry?

answer the question before they ask?

offer them help without them asking?

not say anything to anyone for the whole day?

worry about how much work you're doing?

worry you bought the wrong ink for your pen?

wonder if it's the pen and not you?

think that no one is listening?

change your passwords?

worry about money?

worry that you worry too much?

worry your writing is too messy for anyone to read?

let someone else read what you wrote?

think what you've written is the best thing ever seen?

dance like no one else is watching?

make something just because you can?

try to put the pen down and realize you have a little more to say?

think about only eating vegetables?

think about only eating meat?

worry about what you wrote just a few pages ago?

get in touch with an old friend you haven't seen in years?

try to reconnect?

try to disconnect?

worry your kids will follow in your footsteps?

change the songs on your playlist?

think about how lucky you are?

fill a notebook?

start a new notebook?

think about what's in your other notebooks?

change the size of the words you're writing so it fits in the available space?

prepare for the worst?

get nothing done after spending eight hours at work?

find more important things to do?

wonder if anyone would notice if you weren't here?

make the rules?

think you're explaining so anyone could understand?

have one more than you should?

think about staying up all night reading comic books like

you did when you were a kid?

wash your coffee mug?

wish you hadn't skipped class?

forget what you're writing about?

worry about repeating yourself?

see yourself in other people?

worry you see yourself in other people?

have to edit yourself?

wish you never started overeating?

wish you never started drinking?

worry you're wasting your time?

worry you're wasting someone else's time?

think about what you're doing next?

think about what you're going to say before you walk

over and talk to someone?

let the moment pass by?

think it's just too difficult to stay in touch?

have to choose your words wisely?

worry you've said something out loud, instead of in your

head?

lose control, because you're too excited?

think about where all the dust on your desk comes from?

feel unneeded?

feel unwanted?

imagine what it would be like if you had all the money?

feel like you're not appreciated by the company you work for?

think about what appreciation would look like?

give in to what you want?

not let your anger get the better of you?

think you might be near the end of a project?

wish you hadn't eaten that last double cheeseburger?

try not to listen in on someone else's conversation?

have to pick up and move?

want to punch someone in the face?

resist the urge to run away screaming?

avoid having difficult conversations?

ask too many questions?

not ask enough questions?

know where to find the answer?

know who to ask?

make someone else your priority?

look outside for validation?

visit with your family?

think about how you define your family?

wait?

look before you leap?

run?

go backwards?

say, "thank you"?

28

say, "you're welcome"?

say, "I love you"?

mean it?

feel pride?

feel content?

wish you could start all over again?

try to turn off your mind?

try to ignore your heart?

take a break?

get just one piece of happiness?

actually have fun?

actually enjoy life?

get to do what you enjoy?

think about what you enjoy?

eat too much fried chicken?

get a letter in response to one you mailed to a pen pal?

write, just to write?

write now?

think about other writers?

think about what people would do without you?

worry about getting sick?

appreciate the clear blue sky?

hear information you shouldn't have?

get worried?

want to start over?

meet new people?

avoid people?

go out of your way to avoid people?

read the notebooks you made all your notes in?

wish you were someone else?

think you know the answer?

type out what you write in your notebooks?

worry about getting hacked?

stop and enjoy a cup of coffee?

stop and enjoy the weather?

walk up and down the stairs?

take the stairs, even if you don't want to?

find ways to avoid what you're supposed to be doing?

do what you must now, so you can do what you want later?

think you're a cliché?

get annoyed by everything?

react the complete opposite of what people think you will?

hate people?

feel the need to make something new?

feel the need to rebuild something you previously made?

forget something really important?

forget something, even if you write it down?

wish life had a soundtrack?

let someone else make you feel defeated?

feel unsupported?

feel like writing isn't important?

feel like what you want to do isn't important?

find your balance?

lose your balance?

let someone help you?

ask for help?

close your eyes, take a deep breath, and thank whatever

you believe in?

miss the little things?

throw up a little in your mouth?

like someone less than they like you?

have no idea what you're doing?

fake it?

struggle with what you're dealing with?

think about getting published?

think about not getting out of bed?

get tired of fighting?

waste time hating?

not want to stop, because that next great idea will be the

next thing you write down on this piece of paper?

not want the game to end?

think about how people see you?

worry about how people see you?

wish you had the skills to be in charge?

think it's futile?

resist?

give in?

go with the flow?

go against the grain?

destroy before you create?

look at something and try to come up with ways to make

it better?

look at something and try to make it worse?

buy too many books?

remember to contact your best friend?

realize you missed out?

realize time will go on, no matter how much you try to stop it?

hide what you're writing?

worry about repeating yourself?

run away?

run towards?

wish you could find the perfect pen?

want to want to use a typewriter?

need to…

    …upgrade?

    …downgrade?

eat, even when you don't need to?

wait too long to get a haircut?

worry that what matters to you, doesn't really matter?

wish you could find your bliss?

have trouble reading what you wrote?

think about going bowling?

go bowling?

feel like you drank too much?

drink too much?

feel like you're fighting uphill?

worry too much about what other people...

   ...think?

   ...believe?

   ...hate?

   ...love?

change your style?

worry you don't have style?

worry about what you've given to your kids?

focus too much on what you haven't done?

focus too much on what you think you can't do?

think you're too hard on your…

...kids?

...self?

do you think what someone else wants is more important than what you want?

you think what you need is not important?

not know what you need?

stop what you're doing to do something for someone else?

feel overwhelmed by life?

wish you were better at what you do?

think about doing what you love as your job?

try to explain that games don't matter?

try to just have fun?

have to be the best?

just not care?

have to bribe your kids to do something?

feel bad about bribing your kids to do something?

have to stop yourself from comparing one kid to another?

feel like you've failed?

wish you were nicer?

feel like no one understands you?

get tired of explaining?

get excited about the wrong things?

feel like you're not as good a parent as someone else?

wish you had more will power?

wish someone would be your cheerleader?

feel underappreciated?

feel too old?

feel bad for thinking you're a better parent than someone else?

wish they would…

    …do things on their own?

    …ask you to help them?

fear you're not meant for greatness?

fear you're meant for greatness?

say, "I'm sorry"...

> ...and honestly mean it?

> ...and not mean it?

want to dance with somebody?

have a good idea?

think you have a good idea?

feel like you're...

> ...trying too hard?

> ...not trying hard enough?

worry about quantity over quality?

hide what you're doing?

feel embarrassed?

perform?

act against your nature?

become overly sensitive about something you create?

think you're not good enough?

wonder if you actually know your nature?

stare out the window and wonder if...

>...there's something greater out there?

>...today is the day?

>...you've missed the day?

reject reality?

feel like...

>...the smartest person in the room?

>...you're not the smartest person in the room?

>...you don't belong in the room?

wish you were better at keeping the house clean?

wish you didn't have a temper?

wish your kids would listen to you the first time?

forget how old they are because they're too smart for

their own good?

wish you were stronger?

wish you lived in a different time?

want to stop?

need to stop?

worry you're repeating yourself?

struggle to explain yourself?

let someone down?

imagine you're a super spy?

wish you could take away all of the pain?

think, "I can do that."?

do it?

slow down and take your time?

let your paranoia get the better of you?

say one thing and mean another?

ask them to do as I say not as I do?

get a song stuck in your head?

let someone throw you off your game?

want to drive really fast?

want to ride on a roller coaster?

wish you knew everything?

know you are still learning?

wish you knew your neighbors better?

feel like a nut?

suffer from your mind going blank?

have to start over?

start over, just because it's the simplest way forward?

think about your legacy?

wonder what your legacy will be?

write to avoid doing something else?

do something else to avoid writing?

think about writing a novel?

think of a reason not to?

think about writer's block?

wonder if writer's block is real?

know it's not the pen, but still use a specific pen for each

job?

have to look for a pen?

think you've made too many pens?

show off?

think you're showing off, when you're really just doing

what they asked of you, but you're doing it really well?

look for the easy way out?

race to the bottom of the page only to be overwhelmed

with the emptiness on the next one?

carry lucky charms?

change what you wear as jewelry?

worry no one else can read your handwriting?

wonder what makes you who you are?

start thinking about what ifs?

stop looking back?

wonder if you will ever…

   …grow up?

   …learn?

   …teach?

stop listening to the background?

copy something you created before instead of coming up

with something new?

worry about a pandemic?

think the world isn't fair?

change your priorities?

do the same thing over and over?

try the same thing expecting a different outcome?

change?

pretend to be something else?

think you haven't helped enough?

think there's more you could do?

worry it's too hard to stay in touch?

think you know enough to get in trouble?

have an over abundance of knowledge?

worry no one actually gets you?

worry you're not where you're supposed to be?

let things get to you?

wish some things were a little bit easier?

write in your notebook hoping someone will ask you about it?

write in your notebook hoping they won't?

edit while you're writing?

turn off?

let it go too far?

act inappropriately?

want to just stop writing?

proceed even though you're unsure of the outcome?

stop because you're unsure of the outcome?

realize stopping isn't an option?

create a backstory for someone you've never met?

break a promise?

admit you're wrong?

keep writing just so you don't have to do any editing?

wish your work was appreciated, a little bit more?

think you're getting sick?

stare at the floor while you're walking so other people won't engage with you?

feel like you're not doing enough and then see someone doing way less than you are?

think about what you believe in?

think people don't understand what you do?

try to think of ways out of plans, as soon as you make the plans?

mistake someone being nice for something more?

wish you were the best at something?

wish you had more money?

wish you were better with money?

deflect a tense situation with sarcasm?

run out of ideas?

think you could do better with your first kid?

think it's difficult to deal with someone who is a clone of

you?

feel hollow?

feel hopeless?

feel lonely?

have bad ideas?

wish you were in command?

pledge to stop using social media?

make contact with someone you shouldn't?

pledge to do better next time?

hope there's a next time?

admit you're scared?

have fun?

64

try old ways to have fun?

show who you really are?

rely on someone else to make a decision?

worry you're repeating yourself?

worry you're not real?

enjoy someone else's wonder?

have trouble finding what you're actually looking for?

just mark time?

cobble together a personality?

clean your keyboard at work?

start over?

worry about being trapped in an active volcano?

worry about market volatility?

worry about your own health?

wish someone would give you a hug?

wish they meant it when they ask how you're doing?

go to work when you should take a sick day because

you're worried you'd be wasting the sick day because

nothing is really wrong?

have an irrational need for everything to work the way

it's supposed to?

overreact when everything doesn't work the way it's

supposed to?

not start something because you don't know how it's

going to turn out?

not try?

regret?

wish people would notice you?

wish people would not notice you?

worry about asking for permission?

try to replace someone with someone new?

realize you had been ignoring someone?

run away when you see someone you know in public?

call someone you haven't talked to for a really long time

think it's too late to call someone?

wish you'd write something that didn't sound like a cliché?

feel like you're repeating yourself?

worry about how other people see you?

try not to care about how other people see you?

change who you are, depending on who you're with?

want to eat all the pizzas?

worry once you start you won't be able to stop?

let other drivers get to you?

work beyond your limit?

take on more than you can handle, because you don't

want someone to think you're weak?

wish you were the go to person?

worry you haven't evolved enough?

not know what you need to do to keep evolving?

look for a better way to do something?

wish there was a better way?

had gone to visit more?

could control your temper better?

wish you hadn't skipped class?

complain about things you have no control over?

try too hard to be a hero?

have to force yourself to not involve yourself in an issue,

unless they ask you directly?

underplay what you do?

hold on too tight?

feel the morning on your face and wonder how you made

it this far?

not hold on tight enough?

have the wisdom to know the difference?

wish you knew where the ride was going?

worry people see you as soothing you're not because of

who you're standing next to?

lose track of time?

wish you could be a different person?

understand the story?

wish you understood the story?

look back over what you've written in your notebook?

think about a time you'd like to go back to?

think about a time you'd like to go ahead to?

get lost in the fiction you're enjoying?

become emotionally invested in the fiction you're enjoying?

worry about the fact someone doesn't like you?

pretend like you don't understand?

play it by ear?

try…

   …too hard?

   …to make them like you?

   …to make them stay?

   …to make them remember you?

wish you would stand up for what you believe in?

prep for doomsday?

count the days?

forget to write it down?

wish you knew for sure it wasn't all for nothing?

wish for bliss?

could just sit and drink coffee all day?

wish she already had all the unicorns?

wish it was just a little easier?

have to prove yourself?

want to sound smart?

want to say something stupid?

looked the other way?

write, even though you don't know what to write about?

get trapped in an idea and don't see a way out?

think it's easier to continue than to change what you're

doing?

wish your kids understood you're trying to help them?

appreciate a moment of calm?

doubt you know what you're doing?

wish people liked you more?

wish?

want?

need?

feel?

know?

feel under qualified for anything?

want to punch someone in the face?

look at what you've created in the past?

think you could've done better?

know you could've done better?

look for people you used to know?

get shocked when you see how much people have changed?

offend someone by mistake?

offend someone on purpose?

wish your parents would knock before they walked into the house?

pretend not to listen into someone else's conversation?

appreciate someone else's work?

notice how slow time is passing?

put off starting a new project?

wish people would pay attention to what they're doing?

let someone annoy you?

wait until someone talks to you?

have trouble finding something to do?

not understand people who are bored?

feel like you're not doing enough?

feel inadequate?

get your hair cut?

do something just for you?

resist the urge to run away screaming?

look for ways to help?

under-value someone's contribution?

wish someone would reach out for your hand?

wish people who think they're smarter than you would

realize they're not?

think you have to many notebooks?

think you have too many pens?

wonder if you actually need to be in the meeting?

want to call in and tell them you don't feel well?

feel unnecessary?

feel irrelevant?

ask the wrong question?

just try to do something right?

find when you focus on the writing, it doesn't work?

find something to do other than writing?

feel the need to insert yourself where you aren't

needed?

want to smack some sense into somebody?

see no way forward?

see the only option is back?

look for a way out?

worry about what you said to someone?

think your problems are't as important as someone else's?

hope for a change?

want to know why?

sit and wait for someone to need you?

wait too long?

wait for someone else to make the first move?

do something for yourself without feeling bad about it?

forget who you are?

need to be reminded of who you are?

wish for quiet, and then wish for noise when it comes?

wish you were more brave?

wish you were more decisive?

feel useless?

think you're…

    …in hell?

    …going to hell?

    …leaving hell?

feel like you're screaming the answer at the top of your

lungs and no one is listening?

wish you could relocate?

set goals?

keep goals?

give up on your goals?

get knocked down?

get back up?

stay down?

throw up your hands in surrender?

wait too long to write something down and forget the

original idea?

think the new idea is better than the original?

try to reassure someone even though you aren't sure

about your situation?

start a project, but not finish it?

work on a project because it'll be done in less than a day?

feel good?

feel like you should feel better?

feel proud when someone actually listens to your advice?

lead the meeting even though you don't know what's

going on?

align yourself with people who know more than you?

envy someone because they seem to be more important

than you?

want to stay down?

wish you were the one in charge?

relish the fact you're not in charge?

write even if you don't have anything to say?

write for someone else?

write for yourself?

feel like you're repeating yourself?

think it's time to start editing?

get scared at the prospect of editing what you've written?

think the person telling you what to do, doesn't under-

stand what you do?

wish you were more defiant?

wish you could visit someone?

listen to what someone tells you to do?

get confused by the number of pages you've already used

in a notebook?

think you have too many notebooks?

look to someone else's work to inspire you?

get inspired by someone else's work?

inspire someone else?

have trouble finding a reason?

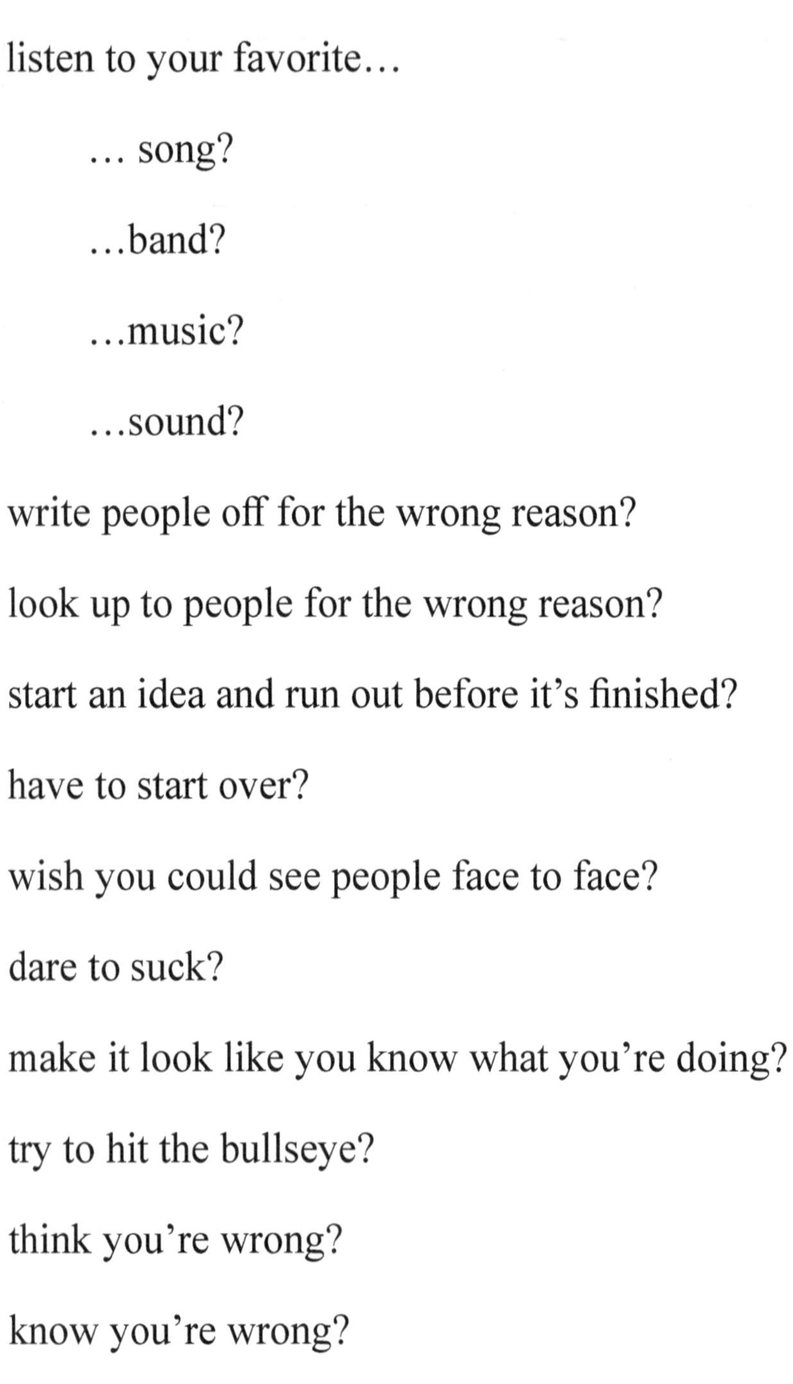

listen to your favorite...

    ... song?

    ...band?

    ...music?

    ...sound?

write people off for the wrong reason?

look up to people for the wrong reason?

start an idea and run out before it's finished?

have to start over?

wish you could see people face to face?

dare to suck?

make it look like you know what you're doing?

try to hit the bullseye?

think you're wrong?

know you're wrong?

worry your proposal wasn't good enough?

think you're going to write something and nothing

shows up?

think the pen is important?

find yourself with actually nothing to do?

wish you were bothered by so many things?

wish magic was real?

forget to write?

wish you were better at your hobby?

think it might be time to start editing?

think writing is like smoking, you never really quit, it's

just longer between puffs?

wish you could be a hero?

feel alone in your own head?

wonder if you've already had your last original thought?

lose the dream/idea when you hold on too tight?

put the process on a pedestal and it gets in the way of the

project?

let the wrong things validate your work?

take the easy way out?

take the path of least resistance?

spend too much time wondering what you could've

done…

   …differently?

   …better?

   …worse?

   …not at all? paint your nails?

let your 7 year old paint your nails?

paint someone else's nails?

realize you need to become reacquainted with someone?

wonder how close you are to destiny?

think about being someone who matters?

actually prepare for a meeting?

feel like you're not working hard enough?

find it easier to write when you should be doing something else?

find lots to do when you should be writing?

stay awake all night, because you can't stop writing?

wish the kids would listen to you because you are actually looking out for their best interest?

wish they would just go to sleep so you could watch one show in peace, all the way through?

take an hour to watch a 30 minute show?

let something interrupt what you're doing?

find the right gift?

wish you had taken more advantage of your time in...

    ...high school?

    ...at college?

    ...at work?

think about how all good people have trouble sleeping?

have faith in who you are?

fear your own shadow?

find the answer and forget to write it down?

get to the end of the page and hope there's enough room

for one more thought, because you know if you get to the

next page the idea is going to run dry?

look for a fight?

avoid a fight?

fight for someone else?

fight for something you believe in?

move around so the guilt can't find you?

start over?

need a break?

wish you could write as well as your favorite writer?

think, "in this moment I am ok"?

want to smack someone upside the head?

deserve it?

take longer than you think you should to finish a project?

lie to yourself?

could you stop eating chips?

wish you knew your neighbors?

think it's ok you don't know your neighbors?

bite the metaphor to death?

forget all you have to do is ask?

wonder about your raison d'etre?

speak with a loud clear voice?

regret?

fit in?

feel like you haven't accomplished anything today?

find yourself bored and unfocused?

think about what it means to accomplish something?

think about what makes you who you are?

think about what you want to be?

say good-bye?

worry about your legacy?

stop…

>…thinking about what if?

>…looking back?

make bad decisions?

act mean to someone when you don't mean to?

say something mean to someone to hide your own insecurity?

wish you were someone else?

have to write something until it sounds right?

think about how…

>…the world should be?

>…you should make it better?

have trouble focusing?

wish you could write the great American novel?

wonder if you're right?

make the wrong decision?

miss the way things were?

put distance between you and someone else when you

don't mean to?

know exactly what someone else is thinking?

get stuck on an idea?

wonder if denying them ice cream one time ruined them'

worry the world won't be able to get back to normal?

worry there isn't a normal because things change all the time'

worry you're not doing what you're supposed to be doing?

have to force it?

keep track of how many days it's been?

want to be alone?

need to be alone?

not want to be alone?

wish you were…

    …someone else?

    …somewhere else?

worry you're a cliché?

lose track of time?

adjust who you are for someone else?

miss your family when you're at work?

miss your work when your with your family?

try to cook something new?

ignore the brief?

wonder when you'll actually be good at your…

    …hobby?

    …job?

    …life?

think about opening your own business?

look…

    …outside for inspiration?

    …inside for inspiration?

miss what's going on around you?

like someone more than they like you?

misunderstand someone?

wish you were cooler?

sit and enjoy the peace and quiet?

wish it wasn't peaceful and quiet?

miss the peace and quiet?

feel like an asshole?

pretend to be nice?

know everyone knows you're a fake?

need a minute to find the groove?

try to find your bliss?

not know what you need?

wish you could find the perfect writing instrument?

think it's the only thing standing in your way?

realize it's not?

get in your own way?

need to unwind?

wish you weren't wound so tight?

see what…

...something could become?

...someone could become?

hope?

mourn?

love?

hate?

win?

lose?

admit defeat?

realized defeat doesn't have to be the end?

get knocked down?

get back up?

look out the window and wonder how we got to this point?

hope someone will read this and not immediately go dark?

wish you were stronger?

believe someone else more than you believe in yourself?

wish…

    …you could finish your screenplay?

    …you believed writing didn't have to be ceremony?

    …believe in yourself?

    …believe in something?

learn from your mistakes?

keep making the same mistakes?

start writing and wonder if you can ever stop?

hope you will never stop?

wish you didn't know yourself so well?

lie to…

    …yourself?

    …someone else?

    …everyone?

hide?

seek?

look back?

write on the walls?

read the writing on the wall?

hide the pain?

wish you were more adventurous?

wish you…

    …didn't have anything tying you to one place?

    …could be honest?

wonder where you will write when this notebook is full?

write the same thing?

get a headache for what seems like no reason?

sit, stare, and think this is not what I

want to be doing?

wish everything would live up to your expectations?

wish staying down was an option?

make a mistake?

get mad at yourself for doing something, while you're

doing it?

wish tomorrow would never come?

wish you could stop

...eating?

...sleeping?

...loving?

...being angry?

...being depressed?

...being unmotivated?

think about getting a tattoo?

take a nap?

wish you could shut off your...

...memory?

...heart?

read comic books?

use sarcasm as a defense mechanism?

smoke too much?

fight for the wrong reason?

    ...right reason?

wish you could draw better?

wish you could you be nicer?

wish you could you be happier?

hide something from your loved ones?

explain things too much?

get worried about your hand getting numb?

feel bad after a dream?

look for people you used to know?

find them?

use a different writing instrument depending on what, where, why, or how you're writing?

wish no one would bother you about what you're writing?

judge how hard you've worked by how dirty your hands are?

judge how well your day went based on the people you talked to?

fall asleep when you don't mean to?

write the same thing over and over so many times it loses

all meaning?

have to force yourself out of bed in the morning?

wish you could stand in the street in protest?

try to figure out what someone's thinking

without asking?

make a beautifully carved bowl, only to have it filled

with plastic unicorns?

hold on to things too long because they might be

important some day?

try to do something nice for someone, only to fail

miserably?

go hours without speaking to another person?

talk to yourself?

figure things out by talking to yourself?

just sit quietly, without feeling bad for not doing something

you should be doing?

wish you were immortal?

pick up a sword?

lay down your sword?

practice like it matters?

look for the truth?

hide from the truth?

wonder if you're out of ideas?

wonder if you're just getting started?

get tired of thinking about it?

try too hard to have it both ways?

have one of those days?

let the dreams from last night affect your day?

stare at a page and pray for just one more good idea?

hope you're not screwing everything up?

miss them, until they're standing right in front of you?

go on even when you don't want to?

lose sight of what's important?

listen to the same song over and over?

wish you had something to say that people wanted to hear?

look ahead?

repeat yourself?

think no one is listening…

   …to you?

   …to the world?

   …to reason?

   …to hope?

change the idea in the middle of a thought?

wish there was more you could do?

laugh?

cry?

smile?

do something easy to get back into the groove?

find it wasn't easy?

wish everyone burst into song?

hope you're as good as you think you are?

ask someone to sit in your office, just so you're not alone?

look at your handwriting and wonder how anyone can ever read it?

write, just to see what happens?

forget to put on deodorant?

wish…

   …looks didn't matter?

   …age didn't matter?

   …money didn't matter?

wish time would hold still so you could master something?

rely on technology too much?

stop using technology and do it the low tech way?

make it look too easy?

wish your job was more challenging?

    …less challenging?

mumble to yourself?

ask whoever is listening, to please let what you were

working on not break?

wish you had a soundtrack?

change, even when you don't want to?

get angry for no reason?

forget…

    …what you were doing?

    …who you are?

wish you could find a better way to

express yourself?

feel like there's nothing else to talk about?

lack the motivation?

find yourself being more frightening than you mean to

be?

finish a letter and realize you weren't out of thoughts?

wish you could meet your hero?

wish you weren't angry all the time?

wish you were motivated to start over because…

…you have to?

…need to?

…want to?

wish you knew who your hero was?

convince your friends and bore your enemies?

drink more than you should?

wish you were a better person?

wish you had something to write on a typewriter?

wish you could go back into the office?

wish you'd made better decisions when you were younger?

wish they would listen to you?

wish you knew what to do?

need to change the question?

just need a minute?

need a safe place?

lose your grip on the moment?

want to drive until you find the end of the road?

have it all figured out?

go dark too quickly?

wish you were something else?

wish you were someone else?

wish you could live a simpler life?

ask questions at the wrong time?

ask tom many questions?

ask the wrong questions?

let the world drive you crazy?

make something just to make it?

keep doing something because you can think of anything

better to do?

put someone in a box?

make silent promises?

forget who you are?

lose a sense of purpose?

have an idea but aren't sure what to do with it?

think you were meant for something more important?

social distance, even if you don't need to?

just sit there and try to figure out where you are?

find something to do other than what you should be

working on?

feel like there's a dark cloud?

want to do something good?

ignore what's going on?

ask enough questions?

ask the same questions?

admit you're better off than you think you are?

think about the decisions lead you here?

wish you could change a decision you made?

worry everyone sees you as a joke?

worry you have too many pens?

wish hadn't sprained your knee?

try to do something only to injure or embarrass yourself?

hide something?

rewatch shows?

rest?

look for new way?

try to avoid making mistakes?

start a project not knowing how to do what needs to be

done to finish?

never say die?

have time to write but no inspiration?

search for inspiration?

actually find inspiration when you look for it?

wish you were one (or two) size(s) smaller?

feel ok with your size?

stand up for yourself?

ask for help?

look forward to the end of the year?

wish you were just a little more outgoing?

wish things were just a little easier?

try to hard to make it fit?

make assumptions?

wish you could do more?

wish more people knew the real you?

wish you were invincible?

waste time?

wish it was another time?

make time?

use all your resources?

hope tomorrow is just a little better than today?

think you waited too long?

think you have a choice?

wish you'd grown up sooner?

have trouble focusing on your work?

wish you did something more fun for work?

stop writing?

become depressed and no one notices?

not want to get up in the morning?

keep moving even when you don't want to?

slow down?

show up?

want to quit?

stare at the clock?

try to think of a good ending?

142

www.ingramcontent.com/pod-product-compliance
Lightning Source LLC
Chambersburg PA
CBHW061733050726
47598CB00002B/468